How the world works:

The Techniques behind the success and the failure of the world

Arthur M. Moore

Table of Contents

Chapter 1
Chapter 2
Chapter 3
Chapter 4
Chapter 5
Chapter 6
Chapter 7
Chapter 8
Chapter 9
Chapter 10
Chapter 11
Chapter 12

Chapter 1

Life is cyclical

  Up and down. That's how you'll be able to describe the real world in 3 words. But rather than accept that life is cyclical, in which dangerous times are normal, we tend to expect that we must always air an upward cycle. Almost everything we do involves alternative people. and since we're emotional and inconsistent beings, outcomes don't seem to be consistent.

Understanding our material world. We can extend that conclusion to life in general. I can't think about an individual's method that's not cyclical. Take personal energy. it'd be nice if our energy would be systematically high, wouldn't it? However, most people have days when they feel great, and when we have days we want a bag of potatoes.

My aim with personal energy is to be as consistent as possible. I'd preferably be at 80% of my full energy potential daily rather than being 95% at some point and 30% the next. There is just such a lot among our management. And even though we work with things that are within our control, we still will control everything. Again, personal energy may be an excellent example. You can have a balanced manner with enough sleep, nourishing food, and regular exercise, however, still, you'll have days you lack energy. Why is that? nobody knows.

 The body and mind don't seem to be like math. However, this is often one thing we tend to not appreciate enough in life. Everything is cyclical. Some businesses grow quickly and bust quickly.

alternative businesses grow slow and ne'er experience any exponential growth. Most jobs become orthogonal at some point. Some days, weeks, or months, you may feel weak. otherwise, you might feel sturdy for years during a row and never get bruised or ill. But nothing can stay similar forever. None of the tops of queries and implications means that you'll be able to predict the future. If you've been feeling low on energy for the past four weeks, it doesn't mean you'll mechanically feel higher next week. It additionally doesn't mean you'll feel worse.

Chapter 2

Understanding meals manufacturing

Food manufacturing influences and is inspired by the aid of using numerous factors.

The Food Market Environment: Availability, pleasantness, safety, and pricing of meals; the comfort of entry; records approximately meals associated with pleasant and dietary cost; and marketing—that is, how sure meals are being promoted on the market and intake.

Natural Resources: Access to land and water; the consequences of weather extrude as they affect land and water.

Health, Water, and Sanitation: Health dangers from manufacturing practices and persistent or seasonal illnesses.

Typically, smallholder families are internet meal buyers— that is, they buy greater meals than they devour from their very own manufacturing. Thus, what's to be had in markets influences their meal purchases and their diet. Smallholders additionally commonly produce a few quantities of meals for family intake to stable get entry to meals they want or prefer. It can be much less high-priced or greater handy to develop a few meals than to shop for them, especially thinking about the fee of ladies' time and the truth that ladies might not usually manage the profits had to gain nutritious meals for their circle of relatives.

To date, biofortification tasks consisting of the ones focusing on the manufacturing of orange-fleshed candy potato have typically centered on the family farming gadget. Over the lengthy term, however, additionally, they aim to grow the availability of greater micronutrient-dense meals in markets. Homestead meal manufacturing sports have additionally illustrated the essential function the family farming gadget performs in choices associated with diets and vitamins. These sports are intended to enhance manufacturing for family intake of fruits, vegetables, legumes, and animal-sourced meals. The proof of fulfillment for those tasks is mixed, however; implementers have stated growth in intake of those nutrient-dense meals, in addition to in advertising of more nutritional diversity, however, the effect on stunting from only domicile meal manufacturing has now no longer been documented. Impact on diets has been more potent in which vitamins education, social and conduct trade (SBC), and ladies' empowerment interventions have been Incorporated.

Many cutting-edge studies, such as a few inside Feed the Future, are measuring the general nutritional effect of interventions that specially goal the family farming gadget. In contrast, few cost chain sports are tracking or measuring the effect on herbal resources, meal availability, and marketplace charges or factors of the fitness environment (e.g., water pleasant or sickness risk). These regions of capacity effect are essential to recollect in designing and enforcing Feed the Future sports.

Through its studying agenda, Feed the Future is devoted to studying with the aid of using, doing, and strengthening the proof base.

Opportunities for Nutrition Linkages

This interest exemplifies how the manufacturing-to-intake and profits-to-buy pathways can interact. Evidence has proven that multiplied manufacturing, profits, or information of vitamins on my own will now no longer affect stunting. PROMASA II included techniques for growing each profit and meal availability and diversity, such as multiple-get entry to animal-supply meals. Crossing each additive is a complete SBC approach centered on family- and network-degree knowledge of maternal, infant, and younger infant feeding and care practices. This layout highlights key possibilities for linking agriculture and vitamins:

Food Production: PROMASA II supported multiplied availability of numerous nutritious meals on the family degree. In addition, it constructed abilities and information at the same time as presenting inputs to assist in enhancing one's circle of relatives incomes. The mixed strategies helped pass beneficiaries along the pathway from meal manufacturing to advanced meals to get entry to and probable to advanced vitamins for kids as well.

Food Market Environment: Even while getting entry to meals is advanced, the supply of numerous, nutritious meals withinside the market is likewise critical. Increased manufacturing can assist stabilize neighborhood marketplace elements and as a result, assist in multiplied intake in goal communities.

Nutrition/Health Knowledge and Norms: PROMASA II labored to grow network information approximately practices critical to accurate vitamins. Community dedication to enhancing entry to animal-supply meals for younger kids is commendable, however, it has to be visible whether or not the brand new social norms—particularly the exercise feeding others' kids—will take hold, and as a result make sure the sustainability of this approach.

Natural Resource Environment: PROMASA II related animal manufacturing to farming practices that enhance soils and decrease the use of chemical fertilizers. However, cognizance of any new farming technique's capacity for detrimental fitness effect (e.g., zoonotic sickness) is likewise required.

Chapter 3

Understanding energy

What is energy? Many people have a preferred concept, however, the nitty-gritty info of energy may be complicated. When it comes right down to completely answering energy-associated questions, it's now no longer unusual to be a piece foggy.

In this book, we'll clean all of it up as we take a deep dive into answering the query of what energy is and all of the info that helps it. Where does our energy come from? Where is the energy saved? What sorts of energy are there? Read directly to examine those solutions and lots extra.

What Is Energy?

In its maximum not unusual place definition, energy is the potential to do paintings. In other words, the whole lot that may do paintings has energy. In the case of energy, doing work is likewise referred to as inflicting or making an alternative. Energy is both converted or transferred each time work is being done. This way that because it modifies paperwork each time it's used, the quantity of energy withinside the universe will for all time stay the same.

Why Is Energy Important?

Why do we want energy? Simply put, without energy, there's no lifestyle. But what mainly makes energy so crucial to our lives? Well, strive to take into account something that doesn't use energy. It's not likely you will give me an answer. A warm drink in a cup, a dozing baby, a bouncing ball, or even a beating coronary heart all have energy.

Nine Reasons Energy Is Important In Our Lives:

1. Breathing

2. Communication

3. Digestion

4. Growing

5. Healing

6. Heat

7. Light

8. Power

9. Travel

The backside line is that lifestyles run on energy. The extra energetic we are, the extra energy we want. Since we can recycle and reuse energy, we must absorb a normal energy stream.

What Types of Energy Are There?

There are principal sorts of energy: kinetic energy and potential energy. Of course, there are numerous exclusive styles of energy, however earlier than we dive into that, let's examine a piece extra approximately those principal classes of sorts of energy.

What Is Kinetic Energy?

Kinetic energy is referred to as the energy of motion. For an item to have kinetic energy, something needs to work on it.

When an aircraft is in flight, an object falls, or the wind blows, it has kinetic energy. The extra mass and the more pace an item has, the extra kinetic strength it has. Kinetic energy is measured in joules (J), the biggest unit of energy.

Types of kinetic energy include:

- Electrical energy

- Motion energy

- Radiant energy (electromagnetic radiation)

- Sound energy

- Thermal energy

What Is Potential Energy?

Potential energy is the saved strength inside an item as a consequence of an item's arrangement, position, or state. A parked vehicle sitting on the pinnacle of a hill and a mild bulb that's grown to become off are examples of potential energy objects.

Types of potential energy include:

Gravitational potential energy

Chemical energy

Mechanical energy

Strong nuclear potential energy

Weak nuclear potential energy

Energy is the ability to do work, however, you want energy to do this work. Luckily, potential energy can convert into kinetic energy, and it works the alternative manner round as well.

What Are the Different Forms of Energy?

We now understand that we classify energy into principal types — kinetic energy and potential energy. But what are the exclusive styles of energy? Well, energy takes on a widespread quantity of various paperwork.

Listed under are the maximum not unusual place styles of energy:

Chemical energy

Electrical energy

Electromagnetic energy

Gravitational energy

Heat energy

Hydro energy

Magnetic energy

Mechanical energy

Nuclear energy

Radiant energy

Solar energy

Sound energy

Thermal energy

Wind energy

Is Light a Form of Energy?

Yes, mild is a shape of energy. More mainly, mild energy is a shape of electromagnetic radiation that we additionally use in microwaves, radio waves, and X-ray machines. We talk about the shape of electromagnetic waves that we will see as mild.

Light energy is charming due to the fact it's miles each the quickest regarded substance withinside the universe and the most effective shape of energy we will see with the human eye. Much like different styles of energy, mild energy also can be converted. For example, photosynthesis takes place while vegetation soaks up mild energy and alternates it into chemical energy.

 Various assets produce exclusive sorts of mild. When warmth assets along with the solar make it mild, we name it incandescent mild. Alternatively, TVs and fireflies are examples of luminescent mild.

What Type of Energy Is Food?

Chemical energy is the energy that is related to meals. When we consume meals, our bodies keep the chemical bonds of atoms and molecules to assist us to live warm, healthy, and energetic. That saved energy is later launched via digestion.

All ingredients offer exclusive quantities of energy, which we degree in calories. When we consume meals, our bodies convert those calories (which can be the saved energy) into chemical energy, permitting us to do work. You can degree the energy you get out of your meals with the aid of counting your calorie consumption.

Most of the energy we utilize from the meals we consume we get in carbohydrates, fats, and proteins. The Dietary Reference Values set with the aid of using the authorities advise that approximately 1/2 of our everyday energy consumption has to be from carbohydrates. We have to then get 20%-35% from fats and the ultimate 10%-35% from protein.

Chapter 4

Everyone needs your hard-earned cash

Managing your finances...
Most people have fought with our mother and father for now no longer shopping for that luxurious phone or for now no longer growing our pocket cash each few months, right? But when you begin to earn income, you can't help but comply with the reality that making ends meet isn't a child's play. You understand positive matters approximately cash (and life) whilst you begin handling your finances. For instance, you're now no longer snug asking your mother and father for monetary assistance and assume two times earlier than spending your difficult-earned cash. In reality, you

arrive at a word whilst you convey domestic cash in preference to taking it away. Here are some matters you may sincerely relate to in case you appear after your finances...

Earning cash isn't a cakewalk

This is one of the few crucial things you understand whilst you begin to earn your cash. Whether you're operating withinside the company quarter or have your business, you need to show yourself withinside the expert international earlier than you anticipate to get the revenue credited for your account each month. In reality, it won't take you long to understand that quite a lot of difficult work is going into income, each unmarried penny, and there aren't any unfastened lunches inside the international.

Savings are crucial!

Since you're answerable for your finances, you all of sudden begin giving significance to saving a few parts of your revenue each month. You pull up your socks to be organized for any cash-associated emergencies and begin maintaining a tab on all of your expenditures.

How do mother and father manipulate it all?

Once you understand how hard it is to earn cash and hold the first-rate lifestyle, you contemplate how your mother and father controlled it during these types of years. They ran the whole family and entertained all of your demands. In reality, they by no means had the posh to surrender from their jobs and take a seat down lower back at home to relax. Why?

Because you were (and are) certainly considered one among their largest responsibilities.

Tax planning

Then there comes a time if you have to come to be an accountable citizen of the United States and pay taxes. You determine the methods to store tax and try to apprehend the meaning of phrases like funding declaration, filings returns, provident fund, shape 16, etc. Not to forget, you usually query how your difficult-earned cash is contributing to the United States' development.

But it's fun...

Not being counted on making money comes with its set of struggles, however, it has its charm. Always desired to head on that solo Europe trip? Now you could as a minimum think about affording it now! Do you continue to dream about proudly owning a Chanel bag? Well, you could chalk out a plan to store cash and purchase it. Once you begin with income, you sincerely come to be a self-structured person and feature the possibility to attain what you've usually aspired for.

Chapter 5

Understanding globalization

  After centuries of technological development and advances in global cooperation, the arena is more related than ever. But how an awful lot has the upward thrust of exchange and the current worldwide economic system helped or harm American businesses, workers, and consumers? Here is a fundamental manual to the monetary facet of this large and plenty debated topic, drawn from modern research.

Globalization is the phrase used to explain the developing interdependence of the arena's economies, cultures, and populations delivered approximately via means of cross-border exchange in items and services, generation, and flows of funding, people, and information. Countries have constructed monetary partnerships to facilitate those actions over many centuries. But the period won a reputation after the Cold War withinside the early 1990s, as those cooperative preparations formed current regular life. This manual makes use of the period extra narrowly to consult global exchange and a number of the funding flows amongst superior economies, in general focusing on the United States.

The wide-ranging consequences of globalization are complicated and politically charged. As with most important technological advances, globalization blessings society as a whole, whilst harming positive groups. Understanding the relative fees and blessings can pave the manner for assuaging issues whilst maintaining the broader payoffs.

THE HISTORY OF GLOBALIZATION IS DRIVEN BY TECHNOLOGY, TRANSPORTATION, AND INTERNATIONAL COOPERATION

Since historical times, human beings have sought foreign places to settle, produce, and trade items enabled via means of upgrades in generation and transportation. But now no longer till the nineteenth century did worldwide integration take off. Following centuries of European colonization and exchange activity, that first "wave" of globalization became propelled via means of steamships, railroads, the telegraph, and different breakthroughs, and additionally via means of growing monetary cooperation among countries. The globalization fashion in the end waned and crashed withinside the disaster of World War I, accompanied by means of postwar protectionism, the Great Depression, and World War II. After World War II withinside the mid-1940s, America led efforts to restore global exchange and funding below negotiated floor rules, beginning the 2nd wave of globalization, which stays ongoing, even though buffeted by means of periodic downturns and mounting political scrutiny.

Chapter 6

Understanding the surroundings

In the context of sustainability technological know-how, the surroundings include interactions among physical, chemical, and organic structures in terms of the organisms inside those structures. In the past, the sector's residing and non-residing structures had been sustained via the surroundings' complicated redistribution of oxygen, nitrogen, carbon, and different elements. However, the dramatic boom of human useful resource exploitation withinside the ultimate century has notably strained the ecosystems we depend upon for survival.

Environmental useful resource control seeks to pick out techniques for stabilizing the currently disruptive courting among human subculture and the residing international. Understanding the surroundings withinside the context of sustainability technological know-how is the relevant distinguishing characteristic of Sustainable Natural Resource Management. Sustainability technological know-how is in an increasing number of famous and vital educational disciplines, designed to cope with complicated real-international troubles via transdisciplinary medical approaches.

Here's how sustainability technological know-how facilitates herbal useful resource control and recognizes environmental complexity.

The Importance of Sustainability Science in a Natural Resource Management Masters

Traditional environmental technological know-how regularly divides unsure factors of the herbal international into identifiable, easy troubles that may be researched and defined in isolation. This specialization permits minute evaluation of conceivable hypotheses, however, is inadequate for information on how those portions of the sector relate to each other as a whole. Sustainability technological know-how seeks to innovatively cope with complicated real-international questions via means of increasing the scope of character medical disciplines to embody transdisciplinary strategies and integrating the numerous stakeholders who might be maximum suffering from those troubles.

Importance of Environment

Environment performs a vital position in wholesome residing and the lifestyles of existence on planet earth. Earth is domestic for specific residing species and all of us are depending on the surroundings for food, air, water, and different needs. Therefore, it's vital for each character to shop and shield our surroundings.

Impact of Human Activities on the Environment

There are specific sorts of human sports that can be immediately attributed to environmental disasters, which include- acid rain, acidification of oceans, alternate withinside climate, deforestation, depletion of an ozone layer, disposal of risky wastes, international warming, overpopulation, pollution, etc.

Chapter 7

Master a skill

Consider any particularly a hit character in records and you'll note all of them have one aspect in common:They're skillful those who realize a way to get stuff done. But here's the kicker:They haven't simply mastered any talents. Instead, they've labored on growing talents that upload cost to their lives. So, that will help you with the same, I am determined to consolidate this list of talents to examine this year. With greater time spent at home now, you haven't any excuse now no longer to examine something beneficial that will help you grow! These talents belong to all spheres of lifestyles, starting from interpersonal to the maximum worthwhile ones. But no matter which sphere of your lifestyle those talents cover, every one of them can extrude your lifestyle for the higher.

1. Speed Reading

Tell me if this sounds familiar: You were given a presentation the following day and you're drowning in an entire stack of documents to examine through. Since there's surely no time to examine all those facts, you decide to skim through to get the gist. The simplest hassle is that as quickly as you begin studying quicker, the textual content stops making sense. Almost every reader has an addiction to subvocalization, which is the manner in which he says the phrases in his head even as he reads them. This addiction is understood to seriously grow during study time. Speed studying facilitates the disposal of this tendency by permitting readers to skim through the textual content and realize it at quicker studying speeds. According to a piece of writing on Forbes, a mean grownup reads at a pace of approximately three hundred phrases consistent with a minute even as pace readers clock in at 1500 phrases consistent with a minute. Imagine all of the time you may shop even studying the paper, getting to know for reports, skimming thru emails, and studying books.

2. The Art of Delegating
This is an ability that I consider everyone, mainly the ones in managerial positions, needs to master. The hassle is that lots of us are looking to micromanage obligations and the humans around us. By getting to know how to delegate, you may lose up to a slow and intellectual area for critical obligations.

3. Learning to Play an Instrument
You may simply remember gambling as a
tool and a laugh ability to examine. But
what if I let you know that getting to
know to play a tool will let you improve
reminiscence and enhance cognitive
functions? It's even proven to retard the
onset of dementia. With such a lot of
intellectual and fitness blessings, that is
sincerely one of the fine innovative
talents to examine.

4. Prioritizing Tasks
Many folks cannot prioritize. But today,
greater than ever, we want to examine
this artwork to grow productiveness and
output. I consider that the largest cause
why humans fail at prioritizing
obligations is their lack of day-by-day
planning. Something as easy as having a
to-do listing so as of precedence will let
you improve your day-by-day output and
effectiveness. But it gets higher...By

getting to know to prioritize, you'll allow plenty of vain obligations that wouldn't make it to the concern listing. And despite that, you'll understand at the cease of the day that you completed greater consequences with lesser effort.

5. Mastering Body Language
According to an analysis posted withinside the Proceedings of the National Academy of Sciences (PNAS), humans remember facts from frame language over facial expressions while perceiving emotions. This look correctly proves that you may get your message throughout to humans greater firmly in case you use the proper frame language. That makes this one of the maximum critical talents to examine. Now, "frame language" itself is pretty much an in-depth problem related to gestures, eye contact, frame posture, and many more.

6. Videography Skills

Have you observed how social media giants like Facebook and Instagram were looking to push greater video content material in your feeds? You'd additionally realize that manufacturers at the moment are robotically advertising and marketing their services and products thru Youtube; lots greater than they used to. There are several motives for that, certainly considered one among that is that video content material has been verified to garner improved person interaction. It's virtually greater, stimulating, gripping, and entertaining. People experience it and need more of it each day. So, through getting to know videography talents like shooting, lighting, and editing, you may journey this social media wave and assist yourself or others generate greater video content material. Oh, and there's constantly the opportunity of creating correct cash out of this ability!

7. Mental Clarity

Did you already know that during 2013, the common character's interest span had reduced to approximately eight seconds? Now, there are a plethora of reasons why our interest spans are losing so rapidly. But I attribute this loss in interest span to our
collective loss in intellectual readability. What you want is to increase the ability to declutter your mind and discover your purpose, drive, and cause. Without a described goal, you're only a hamster jogging in its wheel. And an awful one at that on account that studies suggest that intellectual fatigue extensively influences bodily overall performance as well. So, it's approximately time you lessen your mental fog. By having intellectual readability and described goals, you may

then engineer your mind and determine what you want to learn. With that knowledge, you'll realize what wishes on the spot interest and what can wait.

8. Learning Humor
What if I let you know that being humorous is an ability? I remember it so effectively that I am determined to find it in this list of talents to examine. Being humorous is nearly synonymous with being liked. If you are making humans chortle, then you're mechanically a middle of enchantment for all of your social groups. This brings with it plenty of blessings together with more effect on your friends and improved social exposure. Research posted withinside the American Psychological Association suggests that funny speech is comprehended and retained higher. So, in less difficult terms, humans can apprehend you higher in case you're

humorous. Another look confirmed that humor will increase camaraderie and productiveness in painting environments. Now, as a humorous man myself, the fine recommendation I can provide you with to examine humor is to look at those who are correct at it. Notice how they supply punchlines and cognizance of their timing and tonality. You may also note that pretty plenty of them try and chortle at themselves. Embracing their "weaknesses" is a manner of displaying their self-assurance in themselves. Another awesome manner of getting to know humor is looking at nice sitcoms like F.R.I.E.N.D.S. With time, you'll begin to note yourself cracking comparable jokes because of the characters in those suggestions.

9. Writing
 If you got here to this listing seeking out innovative talents to examine, then

remember this evergreen one. Despite speedy digitization, writing stays one of the most powerful varieties of communication.

10. Learning a New Language
Learning a language is a notable manner of enhancing your intellectual capabilities. When you examine a brand new language, you reveal yourself to a modern set of regulations and vocabulary that work out your mind muscles.

Chapter 8

People will treat you the way others treat you

  Your courting with yourself is the maximum critical one you'll ever foster in your lifestyle. I don't suggest underscoring different relationships, but all of it begins off evolving and ends with you. Everyone desires to be favored and accepted. It is human nature to be healthy and integrated. If you act unconsciously without being aware of your intentions, you're wearing out subconscious commands. If you haven't reconciled those minds, they may dictate your lifestyle. If you put a software program application to run at a selected time of day, it's going to hold to achieve this till you override the function. You are the person of your mind. If you aren't receiving the affection and recognition you deserve, I inspire you to appear inwards and heal the one's mind now no longer in alignment with the relationships you need to attract.

The factor I desire to emphasize is that while you locate peace inside yourself, door occasions will mirror that. Have you met human beings who ate up via means of their victimhood and consider their courting issues to stem from being dealt with badly? One wishes the best track into a TV and inside mins, you'll overhear conversations of human beings being undermined. To draw an easy analogy.

Consider a pointy stone lodged on your shoe even as you stroll round eaten up by means of the ache. No matter who you meet, whether or not now they may no longer be pleasant, your attention is interested in the ache, now no longer your interactions with them. By casting off the stone, you recognize the way it turned into blemishing your interactions with human beings. The ache affected your courting with them as it made it hard to be gifted and engaged. This is what many human beings do. They are blind to wearing unresolved emotional bags and use them as a guard to protect themselves. Yet, the guard does little to defend them however discolor their interplay with others.

Nowadays, I recognize the congruence of a person's phrase and their actions. I check human beings via the means of their deeds and could provide them the gain of doubt on multiple occasions. However, if they undermine my faith, I'm able to stroll far from the courting, no matter whether or not I aim to take advantage of it, financially or otherwise. I preserve no adverse relationships with human beings due to the fact I make it my aim to deal with others as I could deal with myself. And due to the fact I deal with myself in excessive esteem, I foster this behavior in all my relationships. You see it comes right down to values. You ought to be functional for approximately what you cost in your lifestyle. I'm now no longer speaking about career, relationships, or lifestyle reasons but the values you live by.

Chapter 9

Fame and its implications

Fame isn't a brand-new concept. We are seeing our 1/3 or second technology of younger human beings finding 'repute' online. Fame online appears more specific now than it did over a decade ago, nowadays, it's severe business! The aim would possibly nonetheless be to have multiplied fans, however, the number of fans and engagement can now without delay correlate with how much cash you could make.

Anyone and absolutely each person can pick out to make their account a 'business' one. Young human beings are taking part with brands, developing content material, and selling products, and it could be a pretty lot to deal with whilst you're at any such younger age.

Social media repute can carry many nice features however it could additionally probably result in younger human beings relating to themselves in dangerous behavior with a purpose to maintain up with expectancies and 'likes'. In this aspect, it could display the poor lifestyle around repute online and what human beings will do to make cash even if an awful highlight is located on them.

Advantages of repute and the superstar lifestyle

Wealth

Fame and superstar usually include or appeal to and create possibilities to make loads of cash. In a lifestyle wherein your pleasant lifestyle is -all elements held constant- without delay proportional to your financial institution balance, there's something to be stated for a profession that brings with it truckloads of cash. The wealth porn the media topics us to from pricey houses and vehicles to luxurious garments and different materialistic pleasures additionally attracts us in, adults and youngsters alike.

Celebrity remedy

Celebrities get unique remedies anywhere they pass. Who wouldn't need that? Then there's the popularity anywhere they pass and the ego enhancement that includes that and fan adoration. Fame and superstar make certain that you are visible and acknowledged, perhaps even publicly cherished in a lifestyle wherein many human beings sense much less and much less visible.

Negative consequences of repute and superstar

1. Emotional strain and unrealistic expectancies

Fame and superstar grow your visibility. While this visibility can enhance your possibilities to make cash, it could additionally have extraordinarily poor consequences. Being withinside the public eye opens you as much as human beings' feedback, evaluations, and expectancies. All those various evaluations about your lifestyle, your grievance of you, and expectations about your behavior can cause severe emotional strain. Most feedback also is regularly merciless and insensitive due to the anonymity of social media.

Social media specifically is the fertile floor for bullying. Content creators whether or not youngsters or adults should cope with evaluations and grievances approximately the whole lot from their look to how they sound and the whole lot in between. You grow to be in a state of affairs wherein you're continuously pressured to play an element with a purpose to live at the proper facet of factors together with your audience.

Fame is fickle so you can't manage to pay to reduce it to rubble or maintain any opposite perspectives for fear of dropping it all. This strain to preserve a sure ideal public personality can bring about mental situations like tension disorders, consumption disorders, substance use disorders, depression, and continual strain.

2. Commodification and exploitation

The public is more aware of approximately what's occurring withinside the lives of their preferred celebrities than they recognize approximately contemporary affairs. Sure, contemporary affairs may be miserable and boring, however, the media additionally makes a specialty of sensationalism to power visitors to their sites. For them, celebrities and the well-known are commodities to make the most and earnings from.

The media will percentage info whether or not public or personal, irrespective of the poor mild they will solidify the man or woman in, and on occasion even with no consideration approximately the veracity of the story. This is a well-known exercise and one this is assured if one achieves repute and superstar.

3. For youngsters specifically

Fame and superstar come even less difficult for youngsters withinside the age of beginning Instagram debts for newborns. Children regularly no longer consent to have their pix and lives documented and shared on social media. Young youngsters even as they're nonetheless of their lovable years can without difficulty be manipulated and exploited with the aid of using their dad and mom for monetary gain. They are pressured to cope with overt and covert strain to maintain appearing withinside the man or woman or own circle of relatives on social media channels specifically if it's far profitable. There's additionally no manner to recognize how a baby will reply to the pressures of repute and superstar.

Chapter 10

Don't criticize others and everyone's a touch prejudice

  It's very easy to criticize other folks, and then onerous to convey one honest compliment. It's so easy to check yourself during a sensible light-weight and at an equivalent time concentrate on the imperfections of other people.

But criticizing people may be a complete lose-lose scenario that solely creates distance, spreads negative energies, and causes tensions. Criticism is one of the worst types of negative thinking, talking, and acting.

If positive thoughts are artistic thoughts of connecting, including, sharing, and loving, then negative thinking consists of thoughts and words (and consequently actions) that disconnect, exclude and unfold hate.

Since it's not possible to measure a positive life with a negative mind, it's obvious why criticizing others is therefore unproductive and irrational. Therefore, let's put a stop to it.

Why does one like to criticize folks?

On a logical level, we tend to most likely all understand that criticizing people brings no sense to anybody. And however, we still do it. If you are doing it, which means it must bring you some quiet worth or benefit. Well, it will be within the short term. The advantages are emotional, and emotions are most frequently stronger than logic.

That means you want to understand criticizing other people on AN emotional level, to modify it once and for all. Therefore, let's analyze the foremost frequent reasons why we tend to all like to criticize people most and have a tough time resisting it.

You criticize people to form emotional distance

We are often more kind to strangers than we are to our wanted ones. Several couples, oldsters, or siblings are crucial towards each other. most frequently the emotional reason for that's to create distance during a relationship. Criticism may be a good way to show emotion and distance yourself from another person.

Now, why would you wish to try and do that? Well, as a result on the subconscious level you're afraid to be hurt or disappointed. youngsters leave their nests, siblings will be a lot more flourishing than you, your mate could possibly break your heart, and so on.

By criticizing others and that specialize in their imperfections, you'll be able to show emotion shield yourself a minimum of a touch bit (they're depreciation – more about this later).

Understanding that leads us to only one necessary conclusion. It's ridiculous to form distance by criticizing others. By criticizing you are ironically forcing them to harm you sooner or later.

Nobody likes to be criticized, and no one can hurt you more than your negative unbroken mind will. Thus, there's no ought to create distance, solely to boost your thoughts, and feelings of self-worth and switch critiques into praise.

On the opposite hand, generally, we even use criticism to form connections and closeness with other people. that's in cases after we seek for a standard enemy to consequently notice basis with someone we like or can profit from.

But beginning a relationship supported by hate is completely not a decent start. We're only showing off what we are ready to try and do to other people, simply to urge a little bit of attention and love. Negative energies perpetually somehow intensify and backfire.

You criticize people as a result of you envy them

Criticizing others to feel higher regarding yourself and criticizing out of envy are closely connected motives. they're rather completely different tones of an equivalent voice. Let me explain.

It's in our genes to hate unfairness. And once someone gets one thing we wish for in a very unfair way, or after we feel life was unfair to America and type to others, savagely robust feelings of envy arise.

Examples of things that typically build us envious, because life is unfair:

- A friend gets lucky and earns far more money, much a lot simpler than we tend to do

- A parent shows more attention to a sib than to us

- A coworker gets promoted, however, we merit the promotion more

- A colleague is proficient and doesn't
  ought to work therefore onerous to
  be sensible at an exact sport

- We supply far better support to our
  youngsters than we had, but it
  appears they don't appreciate it

- We will notice several similar things

All these situations are unfair. We have a
tendency toll, life will be very unfair
generally which hurts. We tend to shield
ourselves with many various
rationalization mechanisms. We protect
ourselves with self-delusion.

Chapter 11

Understanding the future

The past and also the future is not therefore different. Each past event contains a cause or causes that, as we glance back at them, generally add up to us from our viewpoint within the present. Likewise, every past event has implications and influences the events that follow it. It is a linear sequence that, again, makes sense to us once we relive them. To me, the future works in the very same way. The difference, of course, is that we don't grasp what's going to happen in the future. Instead, we have a wide range of attainable futures. But, even as with a singular event within the past, the events happening today, can form the last word of the future.

Advancements in technology are a region wherever past events will give a decent indicator of how future events can transpire. This is often very true once technological achievements are driven by or impact social trends. As I replicate back in 2017, the technology that I feel will create the foremost future impact is blockchain. The backbone technology behind cryptocurrencies, the distributed nature of blockchain will begin to revolutionize all manner of record-keeping admire contracts, individual health records, and money ledgers over the next five years. It's very like the first stages of the internet, where the underlying technology creates a foundation for all manner of ideas and business opportunities.

Since the dawn of time, folks have tried to predict the future. The fundamental construction of the human brain forces

America to seem forward; this ability is our real (and perhaps only) biological process advantage. In anticipating changes in the environment – what enemies or allies can do, however, the prey will act or threats develop – our species evolved to dominate this planet. In making a model for what we'll react to – droughts, rain, oxen herds, or simply future vacations – we tend to produce our inner worlds.

Futurism as a social issue started with the shamans of constellations and gatherers, who claimed the flexibility to see and listed it for power, or privileges. Then, within Greek culture, a primary conceit to charge views of the long run took place: in powerful supposed tanks called the ORACLES. There are key lessons concerning future thinking to find out from the world's first and most no-hit future prediction business, the

urban center Oracle. Since then folks
from several disciplines have tried to
predict the long run and that we will
learn from their strategies and their
mindsets.

Chapter 12

It's sensible to be humble

What Is Humility?

Humility is the ability to look at yourself accurately as a person with abilities also as flaws whereas being empty vanity and low self-esteem.

Humility isn't perpetually seen as a strength however sometimes it's thought of as a weakness. Some believe that humility has low opinions of yourself, low self-esteem, and a scarcity of confidence. it's after all the opposite, humility is having the concept to grasp that even though you're doing well, you are not compelled to brag or gloat about it.

Benefits of Being Humble

It is vital to remain humble as a result
having humility not solely helps you
develop an additional kind approach to
interacting with others but it conjointly
influences how you understand yourself
and also the world around you.

Strengthens reference to Others

Humility helps one extend more
compassion and sympathy to others.
Those who follow humility are more
likely to think about others' beliefs and
opinions. This is possibly a result of
humility offering the chance to subside
egoism and be more attuned to the
emotions of others.

This is because humility offers the
opportunity to become less self-involved
and more attuned to the feelings of
others.

Broaden Perspective of Self

Humility also helps within the development of self-growth and self-awareness, as a result, it permits one to rationally acknowledge ways in which they will improve themself.

Humility can facilitate developing an additional profound and evolved outlook of the planet and what's happening on it. Humility allows you to consciously bear in mind that you simply bring value to the present world however that several others within the world even have a purpose.

Strengthens affiliation Between Spirituality, faith & Well-Being

Humility could be a spiritual virtue. There is a correlation between humility, positive well-being, religion, and spirituality.

How to Be additional Humble

Let's take a glance at ways in which you'll learn to be more humble.

Don't misunderstand Pride With Prideful

Most would take into account humility the opposite word of pride and will associate pride as being a foul attribute to possess.

Pride isn't a negative thing, it's truly quite important. Pride returns from being proud and there's nothing wrong with being pleased with yourself or wherever you come from.

If someone begins to assume they're higher than others and solely makes choices that rely on what's best for them, they are thought-about egoistic and prideful. neurotic pride makes it tough to be considerate to others or have kind in real relationships. Those who are too prideful may not notice or notice that there are areas within which they will improve.

Do Some Soul Searching

Usually, prideful people show a self-assertiveness that usually stems from unidentified insecurities. too high shallowness isn't truly confidence however after all inhibited negative emotions towards oneself.

Researchers within the field of psychological science found people who displayed egotism and egoism conferred higher levels of "displaced aggression" once hearing insults that vulnerable their egos.

Don't Be a Pushover

Don't confuse humility with compliance. Holding folks in high regard and thought doesn't mean you want to permit them to steer everywhere you go.

You should stand up for yourself and what you suspect and do your best irrespective of what. The purpose of humility is that you simply don't have to be compelled to create somebody who feels no-count while doing so. this does not mean turning into a pushover though.

www.ingramcontent.com/pod-product-compliance
Lightning Source LLC
Chambersburg PA
CBHW051452150726
48000CB00005B/2360